The Alpaca: Its Introduction to Australia

by George Ledger

with an introduction by Jackson Chambers

This work contains material that was originally published in 1861.

This publication is within the Public Domain.

*This edition is reprinted for educational purposes
and in accordance with all applicable Federal Laws.*

Introduction Copyright 2017 by Jackson Chambers

Self Reliance Books

Get more historic titles on animal and stock breeding, gardening and old fashioned skills by visiting us at:

http://selfreliancebooks.blogspot.com/

Introduction

I am pleased to present another title in the "Alpaca" series.

The work is in the Public Domain and is re-printed here in accordance with Federal Laws.

As with all reprinted books of this age that are intended to perfectly reproduce the original edition, considerable pains and effort had to be undertaken to correct fading and sometimes outright damage to existing proofs of this title. At times, this task is quite monumental, requiring an almost total "rebuilding" of some pages from digital proofs of multiple copies. Despite this, imperfections still sometimes exist in the final proof and may detract from the visual appearance of the text.

I hope you enjoy reading this book as much as I enjoyed making it available to readers again.

Jackson Chambers

INTRODUCTION.

In the last pamphlet issued by the Acclimatisation Society, a rapid survey was taken of numerous animals capable of introduction to countries not yet supplied with them, and calculated to add greatly to the prosperity or pleasurable occupation of such countries.

In the present number one animal of that list is dealt with more at large, and one which, from what is already known of it, may be pronounced of inestimable value, and to Australia probably more than to any other country in the world.

The narrative of Mr. LEDGER's spirited, and now successful, enterprise in introducing the Alpaca to our shores, is likely shortly, it is understood, to assume a more complete and elaborate form. In the meantime, the particulars furnished to his brother in London, and by him laid before the Society of Arts, will be read with interest by those principally concerned in the successful result of that great undertaking.

The Alpaca may now be considered permanently established amongst us, and furnishes one notable example of the invaluable additions to our live stock, that, by a judicious system of acclimatisation, may be culled from the enormous list of 140,000 different animals, from which we are assured on the best authority that we may select. In this paper it is predicted that in fifty years an export of nine millions sterling in alpaca wool will be established from Mr. LEDGER's flock alone. But, as tending greatly to hasten the day at which such a result is attained, it is important to know that already a gentleman has arrived in Melbourne, representing an influential commercial house in Peru, which has obtained "concessions" from the Peruvian and Bolivian Governments for the exportation of 1,500 animals, and who is making arrangements to negotiate their introduction to these colonies.

All accounts, to the most recent, are in the highest degree favorable to the project of establishing this animal in Australia. The flock introduced into New South Wales has increased from 283 in 1859 to 417 in 1861. The small herd in the Melbourne gardens has increased from 19 in 1859 to 45 in 1861, and in either flock casualties are almost unknown. The most favorable accounts have also been received of ten introduced in 1859 into Queensland, and referred to in this pamphlet. "No deaths had occurred, and the increase exceeded the most sanguine expectations."

In a recent communication, Mr. LEDGER states that, having several months since depastured a portion of his animals on lucerne, a portion on clover, and a portion on the natural grasses—that fed on the indigenous pasture was unquestionably in the highest and healthiest condition. One of those lately slaughtered for the Great Exhibition contained 22 lbs. of inside fat.

Mr. LEDGER also states that the superior pasture, and more scientific tending of the Alpaca in Australia, produce a finer, larger, stronger, and in every way superior animal to any ever produced in South America; that the staple of wool is excellent, owing to the non-existence of such extraordinary transitions of climate as the animal is subject to in South America—three months of abundant grass, and nine of mere scanty pickings. One fleece sheared in Sydney, of twenty-one months' growth, weighed 26 lbs.

The great adaptability of the Alpaca to Australia is remarked upon owing to its being a *browsing* animal as well as a grazing one. It feeds eagerly on shrubs and trees, and, if deprived altogether of grass, would subsist easily upon the coarser plants constituting our ordinary "scrubs."

The flesh was recently experimented upon in Sydney, and favorable reports were made by the several hundreds who tasted it.

To the successful acclimatisation among us of this animal and its congeners, the Llama and Vicuna, the Society look therefore with the greatest interest. It is impossible to say of what value it may become in after years, not certainly as superseding the sheep or any other stock now depasturing amongst us, but rather as feeding over the sheep's head, eating herbage that nothing else will eat, and from its little use of water being adapted to occupy millions of acres unfit for any other purpose.

Indeed, it is of the utmost importance to recollect that it is quite possible that it may devolve upon Australia to develop this animal to a degree never yet imagined. It has been hitherto virtually monopolised by the South American Indian, one of the most unimproving of all the races of mankind, the most beset by traditionary prejudices and ridiculous superstitions, calculated in every way to cramp the intellect and prevent improvements. It remains to be shown what may be done with animals like these, now to be subjected for the first time to the same treatment that has effected such wonders with the Leicester, Lincoln, or South Down sheep, the short-horn ox, the thorough-bred horse, and other domestic animals long since brought under our control.

THE ALPACA:

ITS INTRODUCTION INTO AUSTRALIA, AND THE PROBABILITIES OF ITS ACCLIMATISATION THERE.

BY GEORGE LEDGER.

I PROPOSE this evening to treat my subject more in a popular and commercial than in a scientific manner, taking more the experiences of practice than the theories founded upon scientific observations, although no one feels more than I do the obligations we owe to them. Having the advantage of constant communications with my brother, Mr. Charles Ledger, who has recently introduced the alpaca into Australia, I will endeavour to place before you the results of his devotion of a great portion of twenty-two years of his life to the study of the peculiar habits of this valuable animal, and to the accomplishment of his enterprise.

It may be asked—Why this interest here in England upon the acclimatisation of a new animal upon lands situate at our antipodes? The answer is, the first physical necessity of mankind is food, the second clothing. Of the principal substances used in the manufacture of clothing—wool, cotton, flax, and silk—wool is second in importance. The acclimatisation of the alpaca simply means the growth of more wool.

There are two great experiments in connection with our national prosperity, which will probably take the lead of all others during the next quarter of a century, viz., the growth of cotton either in India, Africa, Australia, or South America, and the increased production of wool.

The alpaca, the llama, the vicuna, and the guanaco, were unknown to Europeans prior to 1525, when Pizarro, with his followers, first set foot on Peruvian soil; the two former were domesticated, while the two latter, in a wild state, ranged the mountainous tracts of the newly-discovered land. To the new comers, these animals, particularly the llama and the alpaca, appeared to partake of the properties of the camel and the sheep.

I notice the guanaco first, because it is the largest of these animals, but, as it is scarcely to be called a wool-producing animal, the supply being insignificant, and little or none of it being exported, I shall very shortly dismiss it from consideration.

The region between the Andes, or Eastern Chain, and the Cordilleras, or Western Chain, comprises a vast plateau, or rather many table lands, about 12,000 feet above the level of the sea, which is called by the natives "Puna." The surface is principally covered with a short fine grass, and there are many hilly pastures; and here are found the llama, and the alpaca, the guanaco, and the vicuna, but they are not confined to this region. In summer the flocks of llamas and alpacas are driven to an elevation of 15,000 to 16,000 feet, and in winter to an elevation of 12,000 to 14,000.

The guanaco ranges over a greater extent of country than either of the other species, being found on the immense tracts of table land as far south as "Terra del Fuego," and north to the slopes of the towering Chimborazo; in appearance it very much resembles the llama, although nearly half as large again—the wool is much shorter, coarser, and intersected with hair, and worked up by the Patagonians and Aurocano Indians into blankets, ponchos, &c., while its skin is used as a quilt. The meat is the best of the class, and is highly esteemed. It is very seldom domesticated.

The vicuna, like the chamois, inhabits the highest tracts of land; it is the smallest but most graceful animal of either of the species; its flesh is principally used very slightly salted, dried in the sun, frost, and wind, under the name of *charque*, although it is not so much esteemed as the flesh of the guanaco or the llama. Its wool is finer and more valued even than that of the alpaca, but its yield is very small, seldom, if ever, exceeding a pound a year. My brother informs me he has succeeded in producing a cross between the alpaca and the vicuna, which is termed the paco-vicuna, and that its wool partakes of the peculiar softness and superior fineness of the vicuna, and the greater length of staple of the alpaca. I have also heard from gentlemen who have spent many years in Peru, Chili, &c., that the curate of San Antonio, a small town about 40 leagues from Puno, has also succeeded in producing a similar cross, even to the second and third generations, which produce a splendid wool; this reverend gentleman has, I am given to understand, received from the Peruvian Assembly a vote of thanks, and his portrait decorates the walls of the museum of Lima, amongst the the Viceroys of Peru, with an inscription declaring him to have deserved well of his country;

and I believe he enjoys a pension of £120 from the Peruvian Government.

Although the quality of the vicuna wool stands so high in the estimation of manufacturers, the quantity is so small that I have not been able to obtain any record of the amount brought within the circle of commercial operations, a large quantity being consumed in the country in the manufacture of ponchos, sombréros, &c. The skins are brought to this country with the wool upon them.

The llama, the larger and least valued of the domesticated animals, about the size of the red-deer, partakes somewhat of the nature of the Arabian camel. Like the camel, it is used as a beast of burden; like the camel, it can live many days without water, but it is more useful than the camel, inasmuch as its flesh is used for food, and, when young, is savoury and nutritious; and its wool for clothing, and other useful purposes—to a much greater extent than that of its Arabian compeer.

From the earliest periods to which even Peruvian records extend, it appears the llama has been used as a beast of burden; its load is generally from 60lbs. to 100lbs., and 150lbs., and its rate of travelling about three to five leagues a day; it is driven in flocks of sometimes as many as 500 to 1000, and requires little trouble from the drivers, one of the oldest and most experienced leading the way, and the others following.

The organisation both of the llama and the alpaca is admirably suited to the nature of the country they inhabit. The eye, from its size and shape, indicates the possession of a strong and quick sight, and also enables them to bear the reverberation of the rays of the sun from the sand and snow; the sole of the foot is guarded by a cushion, and the toes armed with hard and curved nails, which enable them to climb with ease steep and craggy places; the construction of the mouth and teeth enables them to cut short grass upon the ground, while, joined to the length of neck, and with the aid of the tongue and cleft lip, they can reach and cull herbage growing in interstices of rocks as well as the tender shoots of tall shrubs. The division of the stomach into compartments enables them to retain both food and water, and to use the latter for the assistance of the process of mastication, or to allay thirst.

The number of these animals employed in Peru and Bolivia in the carriage of barrilla, grain, wool, bark, ores, and other products from the interior to the coast, my brother estimates at over 600,000, and for the interior communication between town and town, mines and amalgamating establishments, with ore and fuel, at double that number, giving a total of 1,800,000 so employed.

The llama wool is principally consumed in the country—very little being exported; it is used for sacking in the transport of grain, flour, &c. About 2,000,000 lbs. are annually consumed in this way. The "soga," or cord, by which the load is secured to the back of the animal, will consume annually another 2,000,000 lbs. Many and varied are the articles manufactured from the llama wool—cordage, carpets, bed coverlets, bags and sacks for various purposes, taking perhaps 2,500,000 lbs. The alpaca wool is of greater value and more certain in its demand, so that the llama wool is more used in the country. Those used as beasts of burthen are never shorn, their wool serving the purposes of a pack.

The meat of the llama is highly nutritious; the Indians kill off the old ones of their flocks from time to time during the year.

The Alpaca.

The alpaca stands about 4 feet in height, and is of inferior size to the llama, the size of a full-grown deer, producing a much finer and longer fleece.

The Indians hold for this animal a superstitious reverence, and most firmly believe that any sufferings the animals may undergo, on being driven off their pasture-grounds, will be visited on them and theirs.

In the 16th century, and even from the remotest times, the Peruvians, being comparatively (to the other tribes of the great continent of America) a civilised people, and well acquainted with the arts of spinning and weaving, fabricated from alpaca wool textures of much delicacy and beauty, which were highly prized as articles of dress. And that the use of them had prevailed for centuries, is proved by the opening of very many of the "Huacas" or ancient tombs of the Peruvians, in which the dead had been enwrapped in stuffs made from the fleece of the alpaca. The wool having risen in value, and become an article of so much demand, little or none is at present manufactured in the country, or has been for the last 25 or 30 years. Its fleece is superior to that of the sheep in length and softness, averaging 7 to 9 inches, and sometimes it is procured of an extraordinary length. The fleeces, when annually shorn, range from 7 to 10, 11 and 12 lbs. Contrary to experience in other descriptions of wool, the fibre of the alpaca's fleece acquires strength without coarseness; besides, each filament appears straight, well-formed, and free from crispness, and the quality is more uniform throughout the fleece. There is also a transparency, a glittering brightness enhanced on its passing through the dye-vat. It is also distinguished by softness and elasticity, essential properties in the manufacture of fine goods, being exempt from spiral, curly, and shaggy defects; and it spins easily when treated properly according to the present improved method, and yields an even, strong and true thread.

Notwithstanding the remarkable quality and beauty of the alpaca wool, it was long before its value was appreciated in Europe. According to the best authorities, the first person in England who produced a marketable fabric made from this material was Mr. Benjamin Outram, a scientific manufacturer, of Greetland, near Halifax, who, about the year 1830, surmounted with much difficulty the obstacles encountered in spinning the wool, and eventually produced an article which sold at high prices for ladies' carriage shawls and cloakings; but their value arose more from being rare and curious articles than from intrinsic worth. These were, it is well established, quite destitute of the peculiar gloss and beauty which distinguish the alpaca lustres and fabrics of later times, and after a short period the manufacture was abandoned.

To Mr. Titus Salt, of Bradford, must, undoubtedly, be awarded the high praise of finally overcoming the difficulties of preparing and spinning alpaca wool so as to produce an even and true thread, and, by combining it with cotton warps, which had then (1836) been imported into the trade of Bradford, improving the manufacture, so as to make it one of the staple industries of the kingdom; he has, by an admirable adaptation of machinery, been enabled to work up the material with the ease of ordinary wool.

And now, not only are alpaca goods produced in every conceivable variety and style, but at all prices, to suit the means of all classes of the community. Blended with silk thread, they have the appearance of fine lustrous satin; with figures and patterns thrown upon them in silk of different hues, they serve as admirable substitutes for figured silks, both for ladies' dresses and for waistcoatings, whilst with cotton woven amongst its fibres, the article may be sold at such a moderate price as to bring it within the reach of the most humble.

Gentlemen are provided by means of this fabric with waistcoating as cool as any cotton, yet rich and lustrous as any silk. Dwellers in tropical climates are thankful to possess a black coat which, while it has the appearance of broad-cloth, is not a fourth of its weight.

Most of the alpaca wool taken into the United Kingdom is unshipped at Liverpool, but a small portion is also carried to London. At these two ports, it may be asserted, the whole import is landed. It arrives in small bales, weighing 60, 100, 150 lbs., the first put up for llama carriage, the second for donkey carriage, the third for mule carriage.

Dating from the year 1834, when the importation of alpaca wool sprung up as a permanent branch of commerce, the demand has been a growing one, the quantity imported being in :—

	lbs.		lbs.
1834	5,700	1839	1,325,500
1835	184,400	1840	1,650,000
1836	199,000	1841	1,500,000
1837	385,800	1842	1,443,299
1838	459,300		

Since the year 1843, the returns of alpaca wool imported into Great Britain are of a more reliable character. The following table has been drawn up from data furnished by the Board of Trade :—

	lbs.		lbs.
1843	1,458,032	1852	2,068,594
1844	635,357	1853	2,148,267
1845	1,261,905	1854	1,267,513
1846	1,554,287	1855	1,446,707
1847	1,527,300	1856	2,974,493
1848	1,521,370	1857	2,359,013
1849	1,655,800	1858	2,688,133
1850	1,652,295	1859	2,501,634
1851	2,013,202		

The bulk of these importations have been consumed in England, and the quantity re-shipped for the Continent has been comparatively trifling in amount. We must yet allow some 500,000 lbs. shipped annually from Peru to France and Germany.

In 1836, the price was 8d. per lb. During the last ten years, the price has fluctuated considerably, from 1s. 8d. per lb. to 3s. 9d.

It may be interesting to inquire whether this large supply will be continued, and I regret I am compelled to form the opinion that it will not, unless a change takes place in the manner in which the trade is conducted, unless justice and right are better observed, and unless the Indians (shepherds and owners of flocks) are considered as, and treated as, brother traders, instead of mere producers of raw material. I am fearful the sudden increase of the demand was the cause of greater efforts being made to meet it, which have not been attended by a corresponding effort to increase the number of the animals. I fear the practice of the Indian not to shear the female alpaca has been departed from, and a decrease in the flocks will result. It is true that the attention lately drawn to the value of the wool of the alpaca may also lead to a more intelligent system of cultivation ; indeed, I learn from my brother that a friend of his, who in 1843 did not possess a single alpaca, in 1857 had 15,000 on a large estate held by him.

I will give you an example of how the "trade" is carried on. A party, by some means or other, procures the appointment of "Governador" of a district, and quickly enters into a contract with some mercantile establishment on the coast for a supply of say 500 to 1,000 quintals (100 lbs.) of alpaca wool at 50 dollars (Peruvian) per quintal. As soon as the contract is made, he orders the appearance before him, on the day fixed, of all the "Ylacatas," chiefs or heads of communities, within his jurisdiction; he then apportions the quantity of alpaca wool to be delivered by each, according to the number of alpacas possessed by the community he represents; payment in full is then made in advance, at rates varying from 10 dollars to 15 dollars per quintal. The wool thus collected is tightly pressed by the hands and feet into sacks, weighing 110 lbs., and the Ylacatas are ordered to supply, in the same manner, the requisite number of llamas for the carriage of the wool to its destination. Any resistance on the part of the Indians to supply the wool, and llamas for its carriage, is met by the Governador by imposition of most harassing gratuitous service to the state, such as repairs of roads, foot postmen, domestic servitude; and often by sending the party guilty of desiring to do what he thought fit with his own, as a recruit to the first regiment at hand, with "official" recommendation to the commandant thereof that effectually prevents the Indian for a long time, and often for ever, again entertaining the dangerous and turbulent idea of opposing the wishes of his Governador. Do not suppose that the above is a sketch of an exception to the generality of those in authority over the Peruvian and Bolivian Indian—far, very far, from it; it is, indeed, the rule that actuates the conduct of sub-prefect to turnkey downwards, in all the provinces of the interior of those countries where the Indian is to be met with. If the Indian sees the Ylacata coming towards his hut, and, divining his intent, runs away to hide himself, he does not avoid his persecutor. On his return, he finds the money on the floor, or suspended in a bag from the rafters, with an intimation of the quantity of wool required at 10 to 15 dollars per quintal, and the time of delivery; he cannot help seeing it, and is obliged to take it and supply the wool. If he does not, his alpacas are shorn, and even then, if there is not wool enough to make up the quantity, he is put in prison to force him to pay the deficit at the price contracted to the merchant, and, if this is not paid, his flocks of sheep, alpacas, llamas, &c., are sold to make up the amount.

As a beast of burthen the alpaca is little used, the Peruvian Indians have too much veneration for it, and would consider it a sacrilege so to use it. Europeans, however, occasionally make use of it, and, I believe, in the quicksilver mines of Huancavelica it is found nearly as useful as the llama, although the load it will carry is smaller.

The alpaca was formerly but little used for food by the Peruvian Indians; they seldom, if ever, killed it for the purpose, but would eat it when it died a natural death. Of course, when so obtained, it did not rank high. It has latterly been more used, and, when young, is by some considered delicate and nutritious.

There is still another animal which demands my attention, and which is probably destined to play a considerable part in the future history of wool-producing animals. I mean the cross between the alpaca and the llama, called machurgas, from "machorra," a Spanish word, meaning a barren sheep, of which there are frequent instances in Peru and Bolivia. Walton, in his work, says:—

"From the alliance a beautiful hybrid results, if possible, finer to the eye than either parent, and also more easily trained to work, but, like the mule, it does not procreate;"

in which he is confirmed by General O'Brien, who resided twenty years in Peru, ten of which he served as aide-de-camp to San Martin, the Liberator, a great traveller on the Andes, and a landed proprietor and miner. The General says:—

"There is, however, a beautiful animal produced between the llama and alpaca, much handsomer in form and figure than either, also better adapted for work, but it does not breed. These are the animals I principally used at my mines to bring down the ores from the mountains."

In some parts of Peru and Bolivia, these animals, I am informed, are known by the name of guarisso, which is derived from the Quichua Indian, and signifies a foul thing.

I am not about to enter into a discussion on the vexed question of "fixity of species." I must leave this to be settled by others much more competent to deal with such scientific questions than I profess myself to be, but I must direct their attention to this fact, that the opinions of Walton and O'Brien must now be considered as proved to have been entirely unfounded, my brother having bred animals to the third generation, from female machurgas or guarissos, by reverting back to the original alpaca stock on the male side.

I find mention has been made of yet another animal, the aviru, said to exist in immense numbers in Patagonia; but whether this is a new species, or a variety of the vicuna, has not yet been determined. A rug made from its skins, by the native Patagonians, was exhibited at a meeting of the Literary and Philosophical Society of Liverpool, 18th May, 1857.

It would, indeed, be surprising if animals, so useful to man in every-way as the llama and the alpaca, producing him food, and clothing, both in the shape of skins and wool, and helping him in his labors, should not have created in the conquerers of Peru and Bolivia, and their successors, a desire to transfer them to their own countries. Such has been the case; many have been taken to various parts of Europe—royalty led the van, nobility followed—but, as might have been anticipated, the representatives of commerce have been most active. Time does not permit me to attempt any account of the llamas and alpacas that have been introduced into this country and other parts of Europe, which, notwithstanding all the care bestowed upon them, although they appear to breed, do not appear to have become perfectly acclimatised. They have been regarded as rare and curious animals, fitted for the ornamentation of the parks of our nobility and gentry, than as an article of commerce. I will not here enter into a discusion of whether, under favorable circumstances, the alpaca might not be acclimatised in some parts of this country; I am rather inclined to the opinion that when we know more about it, and its peculiar habits and wants, it perhaps may. I am afraid the main cause why it has not, and perhaps will not, thrive with us, is the humidity of our atmosphere, and dampness of our soil, as well as the unsuitableness of our grasses for its sustenance. Its favorite food in its native country is the ichu or ycho, a rushy kind of grass, of which it is immoderately fond, and which I believe is not found in this country.

Many animals had been exported, and in 1844 the British Consul at Arica was requested to send home sixteen alpacas for her Majesty Queen Victoria, eight of which were shipped on board a vessel-of-war, and eight were brought to Liverpool by the *Octavia*. The day after their embarkation, General Yguain (then Prefect of the department of Tacna) was perfectly furious at their having been placed beyond his reach, and raked up an old decree of 1829, strictly prohibiting the exportation of alpacas under very severe penalties—but which had remained a dead letter, seldom, if ever, put into operation, and the existence of which was almost unknown, and stopped the exportation of a further flock of 40 alpacas intended for Germany, appealing of the Government at Lima to support his act, which resulted in a decree of Congress, April 5, 1845, prohibiting the exportation of alpacas in all the ports of Peru, imposing very severe penalties on those caught infringing the law, which has been strictly enforced up to this day.

This decree has not allayed, as it could not allay, the desire to possess them; opposition often becomes the parent of determination, and when a thing is denied to us we often attach a greater value to it than it deserves, and become more intent on its possession, even when it is not worthy of the efforts we make to obtain it.

The vast success that had attended the introduction of the Merino sheep into our Australian Colonies, and the supposed suitability of its climate and vegetation to the alpaca, led to many attempts being made to introduce them there, but without success. My brother was led to consider a plan, which he ultimately executed, of getting them out of the country, for which he was peculiarly qualified by his long residence (from 1836), his intimate acquaintance with the inhabitants of the interior, and their languages and customs, acquired by having been the representative of some of the first mercantile houses in the country, in making contracts with the Indians for wool, bark, &c., and in frequent journeyings in their superintendence. He writes thus:—

"Several times, from 1845 to 1848, were applications made to me by different parties to get alpacas out of the country, but as I well knew that any such attempt must be attended by difficulties of no ordinary nature, I paid no attention to them. I began at that time to think that this valuable animal required being better known, and its habits studied, hoping that in course of time the decree would be annulled, or some revolution in the country would enable me to get them out through a Peruvian port. In the beginning of 1848 I rented a large estate, Chulluncayani, on the frontier of Peru and Bolivia, and among other occupations, such as collecting wools, copper from Corocoro, and Peruvian bark, commenced breeding the alpaca; little by little I collected at first 200; all sorts of stratagems had I to make use of to obtain these, and then they were old and many infirm ones; every means were devised by the Indians on the estate to prevent my breeding them, but after all, in 1851, I succeeded in being the possessor of more than 600."

In February, 1852, my brother entered into arrangements with a gentleman of Tacna to carry out the undertaking, and immediately started for his estate at Chulluncayani; and after giving directions that in December or January, when the fresh pasturage would begin to appear and render driving them practicable, they should commence their journey towards the frontier of the Argentine Confederation, he returned to Tacna, and went thence to Valparaiso, from which port he embarked on the 24th of December for Port Phillip, *en route* for Sydney, for the purpose of ascertaining, from personal inspection, whether the country into which he was purposing to introduce the alpaca was adapted for its naturalisation. In the following March he landed at Twofold Bay, and, in company with a Peruvian gentleman who had accompanied him, made some excursions for about 12 leagues inland, which satisfied him that "the country was most admirably adapted for the alpaca." On the 22nd of March he arrived at Sydney, where he "became more and more satisfied as to the adaptability of the country for successfully rearing the llama species. Here, as in South America, the climate is dry; it matters not how much rain may fall (even for eight days consecutively), the rarity of the air is not affected by it; there may be, no doubt there is, a difference in the atmosphere immediately before and after heavy rains, but I contend no difference exists in the rarity of the air." On the 21st May he left Sydney, arriving in Valparaiso July 3rd, where he was compelled to seek for a new partner in his enterprise, Mr.—— declining to go on with the matter, and even threatening to divulge the plan to the Government, and succeeded in concluding an arrangement with Messrs. Boardman, Dickson and Co. (with others), for carrying out his project. Peru was then at war with Bolivia, and the attention of the latter Government would naturally be called to its northen frontier, leaving the southern one, through which my brother proposed driving his flock, without troops, and comparatively open to him ; accordingly, without delay, he went to Caldera, and thence to Copiapo, and commenced his preparations for crossing the Cordilleras to the province of Salta, in the Argentine Confederation, and thence into Bolivia, to endeavour personally to learn what had become of the animals he had directed, in the previous September, to be driven in that direction from Chulluncayani.

At Copiapo he again met with a Mr. Samuel W. de Blois, of Halifax, Nova Scotia, a fellow passenger from Sydney, who was desirous of seeing a part of the interior of South America, and he became his companion. Accordingly, on 17th September, 1853, my brother, Mr. De Blois, Pedro Cabrera, their guide, and Pablo Soza, as general servant, with 12 mules and two horses with bells to keep their mules together, left Copiapo, much against the opinion of many friends, it being considered too early to attempt the journey, Pedro, his guide, saying, "Certainly it was early, but with good mules it might be done." After encountering terrific storms of wind, drift snow and sand, and sustaining a loss of one horse and two mules, passing by the remains of a party of fifteen men and thirty mules that had perished two years before, besides frequently being reminded of the dangers of their journey by the skeletons of mules, oxen and donkeys which clearly marked the road, on the 25th they attained the highest ridge, 12,000 feet above the level of the sea, and on the 27th they arrived at Sangil, or St. Gil, the inhabitants of which could hardly believe any person so adventurous as to attempt the passes so early in the season. On 16th October they arrived at Salta, 240 leagues from Copiapo. Salta has a population of 10,000, and the governor of the province resides here.

From Salta, Pedro was dispatched in search of the flock of alpacas, with directions to meet his master at San Christobal, distant 250 leagues, while my brother returned to Molinos with Mr. De Blois, with the intention of seeing him on his road back to Copiapo, and left him at Laguna Blanca, on November 8, 1853. From this date to July, 1858, that of his arrival at Copiapo, with more than 300 llamas, alpacas, vicunas, and their cross products, my brother either procured from the Indians, or bred, great numbers of animals, more than 1,000 of which he succeeded in getting into the territory of the Argentine Confederation; but these numbers were being constantly reduced by the inclemency of the weather (losing 200 of them in one snow-storm), the difficulty of obtaining food, and from his flock having drank of the water of a lake infested with leeches. A second time he was compelled to cross and re-cross the terrible Cordillera to obtain the necessary funds :—

"In January, 1854, it now became necessary for my proceeding to Salta, to receive the money that I expected Messrs. Boardman, Dickson and Co. would have placed there for me. On my arrival some eight days afterwards, I found that they, in lieu of sending funds, had determined to relinquish participation in the speculation. Without hesitation, I immediately determined on proceeding to Valparaiso, and I accomplished the distance—240 leagues, or 720 miles—in nine days. I arrived just in time to catch the steamer passing Caldera for Valparaiso, on the 10th February, and arrived in Valparaiso on the 12th. Since leaving Copiapo, on the 17th September, 1853, I had gone over more than 3,000 miles on mule back. Finding that Messrs. Boardman, Dickson and Co. were determined to have nothing more to do with the speculation, I soon came to terms with Messrs. Waddington, Templeman and Co., for carrying out the undertaking."

It would occupy more time than is allowed to me to trace month by month, or even year by year, the dangers, privations, and vicissitudes he, his shepherds, and his flocks, passed through, during these following four years. They trench on the romantic. His mules and donkeys were frozen to death; two of his shepherds, with their mules, were dashed to pieces, by falling over precipices; he was taken for a political spy, which idea he did not discourage, as it enabled him to keep his true purpose disguised. Twice he was arrested, and once had to defend his flock from forcible seizure. The loss of upwards of 200 of his flock from drinking of the water of a leech-infested lake, in the Calchaquies valleys, compelled him to seek a new spot in which he could habituate his animals by degrees to the kind of food he would be compelled to depend upon for their maintenance during their sea voyage, and also to carry out a plan he had devised of improving the wool of the llama by crossing with the alpaca. This spot was Laguna Blanca, "one of the four valleys which commence from the high table-lands to the west of Tucuman up to the main chain of the Cordillera. It lies in about twenty-seven degrees south latitude, but is nevertheless surrounded by perpetual snow, which crowns the mountain peaks enclosing the valley." Being satisfied that he had at length found a desirable spot for the propagation of the species, not only did he at length acquire a new flock of alpacas, acclimatised to temperatures less pure and rigorous than those in which their predecessors had been reared, but succeeded in educating them to a certain extent for the great voyage which lay before them. He built a hut of stones for himself and people, which furnished an indifferent shelter from the inclemency of the weather. For the animals, large yards were enclosed and provided with troughs, in which they were supplied with their daily rations of dry alfalfa, cut up and mixed with bran, to which they became gradually accustomed. On the arrival of the flock at Copiapo, after almost unexampled hardships, and a separation from his family of six years, in May, 1858, great excitement was excited by their novelty. The city was almost deserted for several days, the people forming an uninterrupted procession between it and Punta Negra (a distance of six miles), where the flocks were at pasture. At length, in July, 1858, 322 animals were shipped at Copiapo, on board the *Salvadora*, of 750 tons.

Respecting the manner in which these animals were got away, my brother says, "In no way have I infringed a single law of either Peru or Bolivia—to break or infringe a law or decree of a Government is one thing, to evade the intention is another thing. The decrees promulgated state that it is illegal to drive alpacas within a distance of 40 leagues of the sea-shore, and the penalty for so doing is the loss of the animals; and every person found with them, and the owners, although not with them, are condemned to ten years labour in chains on the Chincha, or Guano Islands. To smuggle alpacas out of Peru, if successfully accomplished, could be done in twenty days. To avoid being amenable to such law, is why I took such a roundabout way as getting the animals first into the Argentine provinces, and then into Chili for embarkation, in neither of which countries do such restrictions exist."

The fundamental principle of human society, laid down equally by statists and revealed law, is that "the profit of the earth is for all." For the sake of peace—of the settlement of property and society—of political expediency and the comity of nations—that doctrine has, by the common consent of states, undergone modification; but the title of all mankind to inherit the common bounties of the Creator, and the varied gifts of nature, is still the governing principle of political ethics. The Chinese war has by many been deemed sufficiently justified, on the ground that no people are entitled to seal up their territory from the general intercourse of mankind, and to withhold the contribution of their peculiar products from the common stock of human enjoyment.

Of all regions of the earth, that of South America is the most dependent on other countries for useful products. When discovered, it had neither horse, ox, sheep, or pig. All those have been presented to it by the Old World, and Peru requites those gifts by the positive prohibition of the export of its most valuable animal product, thus refusing to mankind a participation in the benefits it is calculated to confer.

The Government of Peru has not imposed a heavy tax on the export of the alpaca. It has not restricted it, confined it within stringent conditions, or regulated the export by irksome customhouse regulations. My brother would willingly have paid any amount of taxation, or complied with any conditions the Government might impose, but it would make no terms, listen to no conditions, hear of no compromise. It insisted on the preservation of a monopoly of an animal whose produce was so useful to man, in violation of that impartiality of commercial intercourse which friendly nations were entitled to expect from its laws. It acted upon the anti-social and anti-mercantile maxim, of appropriating exclusively to itself a gift and blessing intended for the common benefit of mankind. I say, unhesitatingly, that it is a much smaller breach of the strict law of political or commercial morality to evade, or even to break such a law, than it is to make it, and that the obligations of patriotism impelled my brother " to do a right thing by doing a little wrong," in the evasion of a prohibition contrary to the laws of nature, and inimical to the interests of mankind. I may add, that if my brother's act may be called by some, politically or commercially, not strictly moral, he at least sins in very respectable company.

The introduction of one of the earliest flocks of Merino sheep into this country was accomplished under very similar circumstances. George III., in 1787, determined to give them a fair trial, and a few from one flock and a few from another were collected in Estramadura, on the borders of Portugal, and as they could not be shipped from any Spanish port without a licence from the King of Spain, they were driven through Portugal, and from thence conducted to the king's farm at Kew.

On the 28th November, 1858, this flock, consisting of 276 animals, arrived at Sydney, and were immediately landed and temporarily located in the Government Domain. I find, by an official report from my brother to the Secretary for Lands and Public Works, dated 16th April, 1859, that the flock then consisted of :—

    46 Male alpacas, pure breed.
    38 Female alpacas, pure breed.
    110 Female llamas.
    27 Females, cross-bred, between alpaca and llama, in first generation.
    11 Females, cross-bred, between male alpaca and female, from first cross.

    5 Females, cross-bred, between male alpaca and female, from second cross.
  40 Lambs of first, second, and third cross.
    5 Male vicunas.
    1 Female vicuna.
    1 Male cut llama, carrier.

283

All much improved in condition since landing.

After a short sojourn at the town Domain, the flock was sent to Liverpool, about twenty miles from Sydney, until a permanent locality could be fixed upon for them.

The idea of conferring on our Australian Colonies the immense advantages anticipated to follow on the introduction of an animal so valuable as the alpaca, appears to have almost simultaneously presented itself to the minds of several individuals.

Mr. Titus Salt, whose name is and ever will remain so intimately associated with this subject, might naturally be supposed to take a deep interest in it, and finding that the alpacas he had obtained from the late Earl of Derby, and from other sources, did not prosper here so well as he could wish, he sent two small lots to different parts of Australia, thus becoming, I believe, entitled to the distinction of the first introducer of the alpaca into our Australian Colonies. I find one flock of four arrived in the *Marshal Pelissier*, at Adelaide, consigned to a Mr. Haigh, of Port Lincoln, in March, 1857, to whom they were immediately forwarded.

Mr. Haigh writes, 7th Nov., 1860:—"The alpacas are not doing well. They have only increased one; the rest have all died. We lost one the other day, about two years old; it was quite fat, and looked healthy." The number was too small.

In July, 1858, Mr. Salt sent out to Mr. Matthew Moorhouse, Riverton, South Australia, two males and three females. One of each sex died on the passage. The other females have bred, but only one of the lambs has been reared.

A Mr. Eugene Roehn succeded in procuring a flock of llamas and cross-breeds, which he drove overland from Peru to Guayaquil, and thence to Panama. From Panama they were taken to New York, where Mr. Benjamin Whitehead Gee purchased them, and brought them to England, where they were exhibited at Glasgow and Birmingham, and then brought to London, and depastured in a park at Ealing. Ten of these animals were purchased by Mr. George A. Lloyd, of London, and sent in June, 1858, to Sydney, where they arrived 8th November, 1858, and produced £600. This flock was sent on to Moreton Bay. I have not been able to learn how they get on—the much larger and more valuable flock of my brother has perhaps absorbed all interest on the subject. Twenty-three of Mr. Gee's flock were purchased by a committee of Australian merchants (Mr. Edward Wilson, a name now well-known from his efforts in the acclimatisation of animals, birds, and fishes, Mr. Mackinnon, and Mr. Westgarth being among the number), and shipped to Melbourne, where they arrived about February, 1859, and have, up to the present time, gone on excellently. (See Mr. C. Ledger's report, 30th October, 1860.)

While I thus cheerfully and willingly give to those who have laboured in the same field of industry, and yield to them the honour of priority in the introduction of the alpaca in Australia, I think my brother is justly entitled to the honour of being esteemed the largest, the principal importer; I believe there is no record of an attempt at acclimatisation being effected on so large, so stupendous a scale, as that accomplished by him, and at so great a sacrifice of time, labour, and money.

In endeavouring to estimate the capabilities of a country for the successful and profitable maintenance of an animal new to its history, we must look first to its climate, and secondly to the food which it produces, and see if the one is suitable, and the other supplies what is required by the animal for its full development. An animal cannot be regarded as perfectly acclimatised until it is demonstrated that it can live in the locality to which it is introduced, as well as in its native country; that its produce can be turned to useful purposes; and that agriculturists will find their advantage in rearing it on an extensive scale.

I have already informed you that my brother was convinced, by the somewhat hasty visit he made to Sydney, in 1853, of the adaptability of the climate and pasturage of the country to the alpaca, and of this he became more and more impressed while they were depasturing at Liverpool.

In May, 1859, the Government directed him to make a tour of inspection into the interior, with the view of ascertaining the most favorable part of the colony for depasturing the animals. He started in July. His report is now before me :—

"THE SUPERINTENDENT OF ALPACAS TO THE SECRETARY FOR LANDS AND PUBLIC WORKS.

"Liverpool, 23rd August, 1859.

"SIR,—I do myself the honour to lay before you a report of my tour of inspection of a portion of this colony, undertaken by your direction, in search of suitable country for permanently locating the flock of alpacas, llamas, and vicunas.

"In doing so, I beg to state that my own observations of the capabilities of the country, or districts thereof, I visited for affording pasturage to the flock, are confirmed by the opinion of a Peruvian gentleman who, with the sanction of the Government, accompanied me on my tour.

"I started from Liverpool on the 6th July last, and proceeded along the Southern Road as far as Yass, extending my observations of the country over an area of four or five miles, and occasionally a greater distance, on either side of the road. From Yass I directed my course to the Murrumbidgee, whence, via Queanbeyan, Micaliago, and Bredbo, I entered Maneroo. A careful examination of the Bredalbane and Yass Plains convinced me of their suitability to the rearing of the alpacas; the neighbourhood of Micaliago, Bredbo, struck me as no less suitable. My opinion of the adaptation of these places to the above purpose is based chiefly on the marked identity of the natural features of the country with those of that part of South America from which the alpacas came. The country all through Maneroo, indeed, corresponded so exactly with that of Peru and Bolivia, that I could easily believe myself back again in those countries. This similarity was still more apparent with respect to the Snowy Mountains, as that magnificent range appeared clad in their winter garb; Koskiusko reminded us of the Sorata or Yllimani, and, with the Australian Cordilleras in full view, we remembered our trials and hardships among the ranges of their more stupendous and terrible counterparts of South America.

"But it was, of course, on the natural pasturage of these places, as the most important object in our examination, that we bestowed the greatest attention. Not only are the pasturage and herbage, rocks and stones, identical with those of Peru, but I found throughout the districts I have indicated abundance of a description of wiry grass known as the "ichu" of South America. It is upon this grass that the llama tribe mostly feed, being extremely palatable and nourishing, and of which they are immoderately fond. The great importance of furnishing the alpacas with fodder as closely as possible resembling that on which they have been accustomed to feed in their native country, need scarcely be pointed out. It was accordingly my deliberate conviction, and also that of my companion, that the Maneroo district was admirably adapted for the location of the alpacas. Should the Government determine on locating the flock in that district, I would recommend for the purpose the country contiguous to the Snowy River, on account of the facilities which the undulating plains and mountain ranges would afford in obtaining a change of temperature whenever the removal of the flock to a warmer or cooler spot should be desirable. By continual thermometrical observations, I found that a similarity of temperature existed in the months of July and August at Maneroo, to that of the country from which the alpacas were extracted; the thermometer at 7h. 30m., varying from twenty-four to thirty-one degrees.

"The only thing I found to cause any apprehension was the existence of 'fluke' in the sheep; in South America its ravages are counteracted by not allowing sheep, cattle, or llamas to drink the stagnant waters that might be formed from springs, lakes, or ponds; puddles formed by rain are not supposed to cause the disease.

"I beg to recommend that the animals be moved up to Maneroo with as little delay as possible; and as my personal attention is necessary, I beg to request you will think fit to relieve me from my intended journey to New England, at least for the present.—I have, &c., "C. LEDGER.

"The Honourable the Secretary for Lands and Public Works."

The spot selected is called Nimity Bell, between Bombala and Cooma, in the Maneroo district, 260 miles from Sydney.

Before the animals left Liverpool for Arthursleigh, on their way to the locality thus selected, about 200 were shorn in November, 1859. The wool produced was about 6 cwt. of three sorts, alpaca, llama, and cross bred, and valued (by sample) by Messrs. Foster and Sons, of

Bradford, at from 15d. to 2s. 2d. per pound, objection being made to the shortness of the staple.

Mr. Titus Salt bought the bulk, and has favoured me with a report. "Bale No. 1 contains only a few fair average fleeces, the great bulk being too short in the staple (supposed from being clipped too early); some of the fleeces are slightly crossed with llama.

"Bale No. 2 is divided into two divisions: the smaller one seems to be nearly all pure alpaca, but has also been clipped too soon, consequently it is too short for combing."

The prices paid were from 9d. to 18d. per lb., which appears to have been caused by the staple being too short, in consequence of too early clipping. It is to be hoped this will be remedied in future.

In addition to this official report, I find from private letters my brother says:—

"The chain of snow-covered mountains that suddenly presented themselves to my view, on ascending a hill from Cooma, brought most vividly before me remembrances of past privation and hardships endured among the grand and stupendous Cordilleras; and I gazed with delight and enthusiasm on a landscape similar to those my eye had so continually scanned while on my hazardous journeyings through Peru, Bolivia, and Chili."

Many writers on the alpaca, &c., are of opinion that without the ichu grass these animals will not thrive. This grass is found in South America, in Peru, and in Bolivia only; and although the Andes extend from Patagonia to Panama, it is in Peru and Bolivia only that the alpaca is found. From inquiries I have made, I believe this ichu grass is not known in this country, except in botanical collections, and I am led to conclude that the failure of the alpaca with us is mainly to be attributed to this fact, while its presence in Australia, as well as in Peru and Bolivia, justify the anticipation of their acclimatisation in our Australian colonies.

My brother writes to me at various periods:—

"So far the animals are thriving well and augmenting in number.

"The lambs are a most decided success, that is those born in the colony.

"I am on my road to Maneroo with the animals. It will take me nearly two months to drive the flock there—close on 300 miles. I really think that the speculation is a great success (not pecuniary to me). The lambs born here are a great success indeed. Only fancy! I write this from the house of the famous M'Arthur, of Merino sheep celebrity, and the alpacas are grazing in his park, part of the princely 'run' granted to him for introducing the Merino sheep into Australia.

"I am glad to say the animals are thriving wonderfully; of their success I have no doubt.

"I am perfectly convinced that the alpaca will in due time produce immense results to this colony, its acclimatisation and adaptability are no longer problematical, it is undoubtedly an immense success; the flocks are in magnificent order, and thriving wonderfully.

"I am glad to tell you the flocks are thriving admirably. I do not for a moment doubt their complete success.

"We have had more than a month of continued rain, and, although the animals were fully exposed to it, I did not lose one; and at this season of the year they are more susceptible of inclement weather than at any other, on account of the pasturage being more scarce and less nourishing."

The amount of rain which falls on the earth's surface is exceedingly varied, but the moisture of a climate does not wholly depend upon the amount of rain registered by a rain-guage; for some climates are humid, and yet not rainy; others dry, and yet subject to periodical torrents.

My brother writes to a friend as follows:—

"As you are one of the few who feel deeply interested in the success of the alpacas, and sympathise with my enthusiasm, I am sure you will not be wearied of my frequently writing to you about them. I send you specimens of wool, from animals of first cross between llamas and alpacas. It was born on the 27th of April last. It is, therefore, of Arthursleigh growth, and I contend that alpaca wool was never grown at the same rate in Peru. It is truly astonishing. The length of staple and quality are beyond my fondest expectations. The animal yielding it is now little over five months' old, and would now clip fully 7lbs. All are in the same state. The fact is, that in this country we shall soon astonish Peru, and I hope to send fleeces, grown at Arthursleigh, to the next Exhibition in London, that will astonish Europe, too! Send the enclosed specimens to your son in England, and let him show his friends what Arthursleigh is doing."

Since the foregoing was written, I have received the *Sydney Morning Herald*, December 21, 1860, from which I extract:—"Since the arrival of the flock of alpacas at Arthursleigh, the animals have thriven even beyond the most sanguine expectations of Mr. Ledger; so that the ultimate success of the importation is now placed beyond a doubt. The present is the proper season for lambing, and the yield hitherto has been very promising. The number of the flock is now 311, and Mr. Ledger expects that by the end of March next it will have increased to 360. It is Mr. Ledger's intention, in future, to make October the lambing month. In Peru the Indians could not be induced to shear oftener than every other year; the manufacturers having fitted their machinery for a length of staple of two years growth, the practice of the Indians must be adopted here, otherwise our produce would be depreciated on account of the shortness of staple."

My brother took to Melbourne two of his pure male alpacas, two having previously been sent which had been given to the Government of that colony by the Sydney Government.

I will conclude this part of my subject with a part of the Report on the state of the Melbourne flock, which I have already told you was presented to the colony by a committee of gentlemen who organised a subscription here for that purpose:—

"Sir,—I have been delighted to have had the opportunity of personally verifying the statement made to me by my overseer, Pedro Cabrera, on his return to Sydney from this city, as to the splendid condition of your flock of llamas; and I unhesitatingly declare that in their native country it would be impossible to meet with any to surpass, and I very much doubt to equal it.

"I class your stock of llamas as of inferior breed in size of animal, quantity, and quality of fleece.

"By continually crossing the female llama and its female progeny with pure male alpacas, up to the seventh cross, purity of alpaca blood most undoubtedly will be obtained.

"There should not exist a chance of retrogression of breed. Every stage of crossing should be progressive until arriving at the same purity as the male alpacas the Government of New South Wales has forwarded to you.

"I would strenuously recommend the preservation of the flock intact, until such time as every trace or sign of llama blood be eradicated.

"This species of animal requires a dry and pure atmosphere. Humidity under foot does them no harm, unless compelled at night to repose on wet ground.

"I would recommend their being exposed to every vicissitude, changing their folds every now and then during wet weather, so long as they are confined to a limited space for grazing on.

"This animal, when left to itself, at nightfall generally selects a sloping ground for reposing on.

"In my opinion, it would be desirable to confine them as much as possible exclusively to the natural grasses of the country.

"The acclimatisation of the alpaca and llama in Australia is now proved beyond a doubt. The smaller flock in this colony, and the larger one in that of New South Wales, have fully satisfied me as to the adaptability of this peculiar animal to the climate and natural grasses of the country.

"This animal is freer from constitutional diseases than ordinary sheep; less subject to those arising from repletion and exposure to rain. Foot rot, catarrh, and bottle are unknown to them.

"Neither are its young exposed to those accidents liable to befall the lamb of sheep. The mothers are provident and careful nurses, nor do the young ones require any aid to make them suck.

"Except at the rutting season, these animals stand in no need of attention; the shepherd need only visit them occasionally; and such are their gregarious habits, that the members of one flock seldom stray away and mix with others, being kept in a good state of discipline by the old ones, who know their own grounds, and become attached to the place of their nativity, to which they return at night, evincing an astonishing vigilance and sagacity in keeping the young ones together and free from harm.

"By trials, careful study, and intimate knowledge of the alpaca, after an almost daily association with this interesting animal of twenty-two years in South America, and two in Australia, it is placed beyond a doubt in my mind that this animal may be naturalised and made to readily propagate in almost any clime; and every day the facilities and the efficacy of their proper breeding must become more apparent.

"The hardy nature and contented disposition of the alpaca, its extreme docility, and gregarious habits, cause it to adapt itself to almost any soil or situation, provided the air is pure and the heat not too oppressive.

"I had innumerable proofs of its hardiness, and its power to endure cold, heat, damp, confinement, hunger, and thirst—vicissitudes to which it is constantly exposed on its native mountains.

"It is almost superfluous on my part to assure you that at all times I will readily furnish all and every information in my power to give regarding this animal; as also willingly aid by suppling, from time to time, as you may consider necessary, such pure male alpacas as may be required to improve and finally raise your stock to uniformity and purity of blood.

"I will only further add, that the ratio of increase in your flock has far exceeded that in the flock under my charge.—I have the honour to be, Sir, your obedient servant,

"C. LEDGER.

"Melbourne, October 30, 1860."

In endeavouring to estimate what may be the results to our colonies of the introduction of the alpaca, let us look at what has resulted from the introduction of the sheep. In January, 1788, the population of New South Wales was 1,030, and its stock consisted of one bull, three cows, one stallion, three mares, and three colts. (Fairfax.) In 1788 Australia had no sheep of its own, the kangaroo and the dingoe were the only animals of any size that it possessed; and the first taken into the colony were procured from Bengal to provide the colonists with mutton and wool. These animals produced hair rather than wool. They are described in Widdowson's work as possessing "large heads, Roman noses, and slouch ears; they were extremely narrow in the chest; they had plain and narrow shoulders, very high curved backs, a coarse, hairy fleece, and tremendously long legs." By crossing these hair-bearing ewes with an Irish ram, Captain Macarthur effected great improvement, and he was persuaded that the introduction of the Merino sheep into the colony would be of the utmost consequence, and in 1797 succeeded, with the aid of Captain Waterhouse, of H.M. Navy, in procuring a small flock of three rams and five ewes from the Cape of Good Hope, originally brought from Holland, which he had the satisfaction of seeing rapidly increase, their fleece augment in weight, and the wool very visibly improve in quality. He crossed all the mixed-bred ewes, of which his flock had previously consisted, with the Merino rams. The lambs produced from this cross were much improved; but the produce from the second cross far exceeded his most sanguine expectations. He expressed the opinion that in the fourth cross no distinction would be perceptible between the pure and the mixed breed. In 1796 the public and private stock of sheep in the colony amounted to 1,531; in 1801, to 6,757, which is 633 over and beyond a calculation of Captain Macarthur, on the basis that they would double themselves in two and a-half years.

The following is a return of live stock in the colony of New South Wales, from the year 1848 to 1857, inclusive :—

Year.	Horses.	Horned Cattle.	Pigs.	Sheep.
1848	97,400	1,366,164	65,216	6,530,542
1849	105,126	1,463,651	52,902	6,784,494
1850	111,458	1,374,968	52,371	7,092,209
1851	116,397	1,375,257	65,510	7,396,895
1852	123,404	1,495,984	78,559	7,707,917
1853	139,765	1,552,285	71,395	7,929,708
1854	148,851	1,576,750	63,255	8,144,119
1855	158,159	1,858,407	68,091	8,602,499
1856	168,929	2,023,418	105,998	7,736,323
1857	180,053	2,148,664	109,166	8,139,162
1859	200,713	2,110,604	92,843	7,581,762

As these non-indigenous animals have thriven here so wonderfully, as they have also done in South America, and there appears to be considerable similarity between the two countries in temperature, climate, mountainous elevation, and natural pasturage, am I not justified in anticipating a glorious future for the alpaca in Australia?

The following is the quantity of wool imported into the United Kingdom from all our Australian colonies:—

	lbs.		lbs.
1820	99,415	1841	12,399,362
1821	175,433	1842	12,979,856
1822	138,498	1843	17,433,780
1823	477,261	1844	17,602,247
1824	382,907	1845	24,177,317
1825	323,995	1846	21,789,346
1826	1,106,302	1847	26,056,815
1827	512,758	1848	30,034,567
1828	1,574,186	1849	35,879,171
1829	1,838,642	1850	39,018,221
1830	1,967,309	1851	41,810,117
1831	2,493,337	1852	43,197,301
1832	2,377,057	1853	47,076,010
1833	3,516,869	1854	47,489,650
1834	3,558,091	1855	49,142,306
1835	4,201,301	1856	52,052,139
1836	4,996,645	1857	49,209,655
1837	7,060,525	1858	51,104,560
1838	7,837,423	1859	53,700,542
1839	10,128,774	1860	55,270,776
1840	9,721,243		To 31st October.

I have already shown you how the exportation of alpaca wool from South America has of late years increased; I now give the following calculation, made on the probable growth of our alpaca flocks in fifty years—a long time in the life of a man, a short period in the history of a people:—

We commence in 1861 with 200 females, 50 males=250.

Females.	Lambs would yield	Males.	Females.				TOTAL.		December.
							Males.	Femls.	
200	120	60	60	at 60 ⅌ ct. (allowing 10 ⅌ ct. for deaths)			110	260	1861
200	120	60	60	,, ,, Those dropt last year will not lamb			170	320	1862
280	160	80	80	,, ,, The female lambs 1861 will drop this			250	400	1863
340	200	100	100	,, ,, ,, ,, 1862 ,,			350	500	1864
420	250	125	125	,, ,, ,, ,, 1863 ,,			475	625	1865
520	260	130	130	at 50 ⅌ ct. only ,, 1864 ,,			605	775	1866
645	320	160	160	,, ,, ,, ,, 1865 ,,			765	935	1867
775	387	190	190	,, ,, ,, ,, 1866 ,,			955	1165	1868
935	467	235	235	,, ,, ,, ,, 1867 ,,			1195	1400	1869
1322	661	330	330	,, ,, ,, ,, 1868 ,,			1520	1730	1870

There will be, after deduction made for wear and tear, accidents, &c., 3250, as per above calculation. I further deduct 25 per cent.

of total every period of ten years, thus leaving in round numbers 2500; at same rate, in

20 years there would be		20,000
30 ,,	,,	160,000
40 ,,	,,	1,280,000
50 ,,	,,	9,760,000

At 7 lbs. wool—each 68,320,000 lbs., at 2s per lb., £6,832,000!

From this it will be seen that, making deductions of a liberal nature, according to the present ratio of increase, there will be in 50 years 9,760,000 head, the wool of which, at 2s. per lb., will amount to the sum of £6,832,000 per annum.

When figures like these are given, incredulity is naturally awakened; but I do not know that there is anything unreasonable in the calculation. At all events, any reasonable reduction may be made, and still leave a value sufficient to deserve the energy and solicitude of the public.

This is not my own calculation, I take it from the *Sydney Herald* of August 3, 1860, and my brother thus writes respecting it :—

"SIR,—In your edition of 3rd instant I have read an article on the probable result, fifty years hence, of the alpacas, llamas, &c., introduced by me into this colony, in November, 1858. I see nothing improbable, as to such anticipations being realised; on the contrary, my experience of this animal, in South America and in this country, fully warrants the estimates, referred to as being effected every ten years, being carried out.

" Two hundred breeding females, and fifty males, produce, as per said calculation, 3,250 in ten years, or equivalent to multiplying original stock by twelve; continuing at the same ratio, the second period would bring 39,000. Now, instead of following up at that rate, and so as to make all and every allowance for unforseen contingencies of epidemics, bad seasons, &c., I deduct furthermore one-third, and multiply by eight instead of twelve, still reducing the 20 per cent. periodically ten years, giving, at the end of fifty years, 5,606,720 animals, which, at 7lbs. wool only—39,247,040, at 2s. per lb.—£4,924,704.

" The figures are large, no doubt; the time, too, is long. I do not wish to appear a visionary Utopian, although an ardent enthusiast, and hope I may not, through excess thereof, have been led into exaggeration. Figures, something like the above, I worked out nine years ago; they often appeared before me—in my mind's eye—during my solitary journeys, and more than once urged me to persevere, when ' to hope seemed hopeless.' "

It will be borne in mind, that while the sheep has increased in the manner I have shown you, mutton has not been an article of food prohibited at the tables of our antipodean relatives; indeed, at the period of the immense influx of a new population, tempted by the recent discoveries of gold, fears were entertained that the appetites of the diggers, joined to the desertion of the flocks, might act prejudicially on the interests of the wool trade; time, however, has proved this alarming anticipation to have been unfounded.

Notwithstanding the enormous draught constantly made on the flocks to supply the daily demands for food, notwithstanding whole flocks were consigned to the boilers by their panic-stricken owners, notwithstanding disease, caused and rendered more destructive by desertion, swept away large numbers, they are not now diminished, but show a steady increase.

A fortunate climate, and an intelligent devotion to the rearing of sheep, has prevented so great a calamity. Regions long thought barren, are now showing abundant pasturage. Irrigation, hitherto unthought of, has supplied, and in the future will supply, the only deficiency of which the country has to complain. The alpaca, living to the age of fourteen to sixteen years, and not, like the sheep, having daily demands upon its numbers for the purposes of food, is more likely to fulfil the calculations I have given to you. By feeding on a coarser pasturage than the sheep, it will benefit the owner of land by bringing into use portions hitherto unproductive. It will bring more capital into operation. Labour will become employed in the new product—the shepherd tending the flocks, the sailor in transporting it to the seat of manufacture, the spinner and weaver in forming it into the beautiful fabrics that spring from the looms of Bradford and Saltaire. The ship-owner and the merchant also will reap a profit, while the wool is passing through their hands or is under their charge, and various classes of labourers will gain a portion of their means of existence in passing the wool from place to place, and from hand to hand in the various phases it must pass through from the raw state until it is displayed as clothing on our persons.

The following discussion ensued :—

The Chairman, Wm. Haines, Esq., said it was now his duty to ask gentlemen present, who possessed any information upon this subject, to discuss the various topics suggested by the interesting paper they had just heard. There were many points which were worthy of notice :—first of all, the energy of an individual gentleman who, under great difficulties, introduced a new species of animal into our Australian colonies; secondly, the importance of promoting an increased production of wool when our supply of cotton might be in danger; and thirdly, the necessity which the present state of events imposed upon us of encouraging, by every means in our power, the production and importation into this country of the largest possible supply of raw material of every kind. He believed there were gentlemen present who were qualified to give them information upon the wool trade and the fitness of the alpaca for Australia. There was one gentleman in particular to whom reference was made in the paper, and he was sure the meeting would be glad to hear the observations of Mr. Macarthur.

Mr. MACARTHUR had no information to convey with respect to the alpaca, for he was sorry to say he knew nothing of the habits of that animal beyond what might be acquired by cursory reading. Before he left New South Wales he had the opportunity of once or twice seeing the flock of alpacas introduced by Mr. Ledger, and he had no hesitation in saying that the paper read that evening conveyed a mass of most interesting information on the subject. The period during which Mr. Ledger had devoted himself to this object was somewhere about twenty-two years, and for nine years he was separated from his family in pursuing this enterprise. He mentioned this fact in order that the meeting might appreciate the great exertions which had been made by Mr. Ledger in accomplishing his object. A vast sum of money was requisite for this purpose; he believed the amount was not less than between £4,000 and £5,000, which was the value put upon the flock at Sydney, and he knew that the amount occasioned difficulty in forming a company to undertake the purchase of the flock and the production of wool. It was considered too large a speculation, and required too much money for the settlers in a young colony to embark in it. The Government, therefore, he thought very properly, stepped in and assisted Mr. Ledger under those circumstances, for there could be no doubt this was just one of those occasions when the Government might intervene to supply the place of individual enterprise. It was hardly to be expected that two or three individuals, or even a company of persons, should embark upon a speculation of this kind, and, moreover, Mr. Ledger was then comparatively a stranger in the country; but the Government having thoroughly investigated the subject, determined to interpose and make arrangements with Mr. Ledger for the accomplishment of the important object he had introduced to their notice. He was glad to see, from the papers which arrived a few days since, that there were strong probabilities of some advantageous arrangement being made by which the Government would divest itself of the property in the alpacas and make them over once more to Mr. Ledger himself. No doubt that was the most advantageous course that could be adopted, as it was not a matter in which the Government should engage itself, unless under the exceptional circumstances which had been stated. The suitability of Australia for a wool growing country had been established by the figures relating to that commodity, which had been given in the paper, and did not require any corroboration from him. There was one fact which he had noted as very remarkable— that was the statement contained in a letter from Mr. Ledger, from Arthursleigh, as to the great length and fineness of the alpaca wool produced in Australia, as compared with that of the animal in its native country. That agreed with the characteristic of the Merino wool produced in Australia, which was remarkable for its great length and strength, as well as the fineness of the staple. He recollected that, some thirty years ago, it was a subject of complaint amongst the woollen manufacturers, that the wool of Australia was of too short a staple to be of much use to them for cloth; and it was then applied to other articles of manufacture, particularly mousselines-de-laine, which were principally made of Australian wool. The wool, however, now was of a much longer staple, and was particularly noticeable for its softness, especially the Merino wool. He would not attempt to enter into a calculation of the vast extent of country in Australia that was suitable for pasturage, both of sheep and llamas. He thought this country had reason to congratulate itself upon the fact that Australia was likely to produce a very large supply, not only of sheep's wool, but also of that description of fleece which had been stated to be so valuable for other classes of manufactures, and there was no saying to what purposes the genius of our British manufacturers might not apply the wool of the alpaca, when it came into the market in sufficient quantity to make it worth their while to turn their attention more especially to it. Referring again to the pasturage capabilities of Australia, Mr. Macarthur alluded to the communications recently made to the Geographical Society by Mr. Stuart, showing that the interior of the country was not, as was previously supposed, a desert. This had always been his own opinion, and he was happy to find it corroborated by so eminent an explorer as Mr. Stuart. Whilst the western portion of the country varied in elevation from 1,100 to as high as 7,000 or 8,000 feet, the average elevation of the table land was not more than from 1,500 to 2,000 feet, which ensured a very temperate climate; and the late Sir Thomas Mitchell, who was a great explorer of the interior, in his work upon tropical Australia, spoke of having experienced nights of intense cold within that tropical region; so that the variations of climate which that country exhibited, might be considered as affording all the essentials requisite for the successful growth of the llama wool, and the same remark was equally applicable to cotton, some very fine specimens of that commodity having been already produced in Australia.

Mr. F. T. BUCKLAND expressed his high gratification at the information conveyed by the paper, and also at the magnificent specimens of wool upon the table. He hoped they would not rest contented with introducing the alpaca into Australia alone. He thought they ought to try the experiment in this country, and when they saw the beautiful garments which were produced from this wool, he was sure they would have the aid of the ladies in bringing such animals into England. He thought that both the alpaca and the llama would thrive well in this country. They had been told that very fine wool had been obtained from the animals in the possession of Miss Burdett Coutts, which were living not more than three miles from London; and his friend, Mr. Waterhouse Hawkins, could tell them that the animals did well in Lord Derby's park, in the north of England.

Mr. P. L. SIMMONDS said that at the outset of this paper, Mr. Ledger had observed that probably it might be asked what interest had this country in a question of seemingly local interest, like the introduction of the alpaca into Australia? To this he (Mr. Simmonds) would respond, that Great Britain, as a manufacturing country, had the greatest interest in promoting the extended production of wool in all countries, and more especially in her own colonies. The wool manufacture—the second of the great manufacturing interests—which engaged a capital at the present time for raw material, labour, machinery, and value of goods made, of fully £40,000,000, was, like the other great textile industries, insufficiently supplied with raw material for the enhanced demands made upon it for home consumption and export. Large as had been the increased production of wool in our African and Australian colonies, yet, with the competition from Continental buyers, we were stinted in supplies, and had to pay enhanced prices for what we did get. And, what was worse, our manufacturers were driven to the use of ragwool, or shoddy, to the extent of 50,000,000 lbs. a year. To meet the demand last year, with a deficient home clip, and with largely increased exports of home and colonial wools, the deficiency in supply for our woollen manufacturers became very apparent. Any increased supplies of wool for the present or future would be of the greatest benefit to the kingdom. Hence he looked with hopeful interest to the efforts of Mr. Ledger in Australia, and to the information which had been laid before them that evening as to the probable results of acclimatising the alpaca there. As the meeting had just heard, it was only a quarter of a century ago that alpaca wool was first introduced into this country, and for the first five years the average imports were only 560,000 lbs. In the last five years the imports had averaged 2,600,000 lbs. per annum, and the advance in price in this period had been from 10d. to 2s. 6d. per lb. Constant as was the demand for this valuable long wool, which had been the making of Bradford, the supply had been almost stationary for the last five or six years, and, Mr. Ledger had told them, would probably decrease instead of increase, and in that case where was Saltaire to find a substitute? They had heard that evening that the alpaca had now been introduced into the three principal Australian colonies, and with every prospect of their doing well. For his part, looking at the wide extent of the country, the varied climate and temperature, the elevated regions that were to be met with in different localities, from Queensland in the north to Victoria in the south, that the pasturage agreed with them, and that even their tall, favorite grass or reed, the *ichu*, was found indigenous in New South Wales, he saw no reason why they should not prove a success. At all events, the accounts they had heard, both from Mr. Ledger and Mr. Macarthur, went to disprove the opinion so confidently advanced on a previous evening, that Australia was totally unfitted in every respect for the alpaca. Many years ago he (Mr. Simmonds) had advocated in his *Colonial Magazine*, and other publications he was connected with, the introduction of the camel and the alpaca into Australia. Had the camel been earlier introduced, they might not now have had to mourn the loss of Dr. Leichhardt and other enterprising explorers who had lost their lives in penetrating the great interior of that continent. He hoped also to see soon the vicuna and the guanaco introduced into Australia, for, although less valuable as wool-bearing animals, yet they might aid the supply of food hereafter, and fill the place of the kangaroo, which was being fast exterminated. These animals, which ranged in such numbers from La Plata and Chili almost down to Cape Horn, would require no care, but would find abundance of food and suitable localities in Australia. It was even possible that the alpaca might be successfully introduced and naturalized in many other parts of the British possessions, such as Tasmania and New Zealand, parts of the Cape Colony, Natal and India, and in Vancouver and British Columbia. But these were matters for future consideration. There might, and doubtless would, be failures, but useful enterprises, because they were new and apparently difficult, should not be discouraged or opposed by either sarcasms or sneers. When it was remembered how many animals, natives of tropical countries, were even now kept in health in so changeable a climate as Great Britain, there was hopeful encouragement for experiments under more congenial latitudes. Such efforts as those of Mr. Ledger were calculated to be highly beneficial to the colonies and to the mother country, and to stimulate others to exertion and enterprise in a similar direction.

Mr. B. W. GEE said his name having been mentioned with some prominence in the paper, he would offer a few remarks. A residence of many years in Australia enabled him fully to confirm the statements they had heard that evening, as to the adaptability of that country for the introduction of animals of the llama genus. Some eight or nine years ago some gentlemen in Sydney subscribed a sum of money for the purpose of sending agents into Peru to obtain a stock of alpacas, but they returned from the mission without having effected the object—they did not obtain a single animal. It had been attempted by other persons, but it was left to Mr. Ledger to achieve success; and although, perhaps, some of his (Mr. Gee's) animals might have been the first to land in the colony, still to Mr. Ledger was entirely due the merit of being the first successful introducer of this animal into Australia. It might be interesting to some to hear a few particulars of the way in which he (Mr. Gee) obtained possession of the flock of alpacas with which his name was associated. Upon his return from Australia, about three years ago, he took it into his head to visit New York, and his arrival in that city was nearly contemporaneous with that of a flock of llamas, which had travelled a distance of about 4,000 miles on foot, having crossed the isthmus of Panama during the hottest weather. These animals were advertised for sale, having been previously exhibited at the Crystal Palace, in New York. The poor animals, from the long distance they had travelled, were in a very bad condition at the time they were brought into the market. The proprietor of the flock, who was a Frenchman, had the modesty to ask £100 each for them; but at that time dollars were very scarce in New York, as it was during the monetary panic. They remained on hand till the winter, when they were put out to grass; and with reference to the question how far the animals could stand varieties of climate, he could say that he saw them nearly up to their backs in snow, with scarcely anything to eat, on Manhattan Island, where there was scarcely anything but stones. After having passed the winter in those most inhospitable quarters, the flock was advertised for sale in New York in the spring. He would read a short extract from the newspaper report of the sale:—

"The thirty-eight llamas that were imported into this city last fall from Peru (or Chili), were offered at auction on Saturday, March 20, at the 'Dyckman Farm'—a farm of four or five hundred acres of land, in the City of New York. It is situated on the Harlem River, below King's Bridge. It is principally occupied as a grazing farm for bullocks 'left over,' or waiting for market at the Great Bull's Head, in Forty-fourth street—the proprietor of the sale-yards having leased it for that purpose. It was in consequence of this occupancy that the llamas were sent up there to winter, they having been taken when landed to the market-place yards for keeping and for sale.

"It seems as though a chain of misfortunes has attended the first attempt to introduce a breed of domestic animals into this country—discouraging, we fear, to all future efforts to add a new, and, we doubt not, a profitable class to our present stock. If we are not mistaken, the shipment was made from a Chilian port (we understand exportation of llamas is prohibited from Peru), by steamer to Panama, and consisted of seventy-two head. They were detained some three weeks at Panama, awaiting a vessel at Aspinwall for New York; and although in charge of a native shepherd, eighteen or twenty of the flock fell victims to Panama snakes, scorpions, poison herbage, and other Isthmus casualties, in the hottest part of the season. The remainder were then brought over in the cars and shipped upon a brig too small to afford comfortable accommodation, with a bad provision of food, and therefore it is not a wonder that only forty-two of the number reached New York alive. It is a wonder that all did not die, and that only four of the weakest lambs died after they were landed, since the whole of them were in such miserable plight that it was thought unwise to offer them for sale. They have, however, wintered better than a flock of sheep would, if landed in the same condition, and all appear now very lively and healthy, notwithstanding their unwonted and long feeding upon dry forage; and, as an experiment, this has proved that these animals are easily wintered in this latitude, and that they prefer the coarsest herbage, either green or dry. In Chili they are fed upon alfalfa, a very coarse kind of clover, and they might, if domesticated here, be fed the same, or on pea-vines, bean-stalks, buckwheat straws, or coarse weeds—such as our animals reject.

"If adopted into our family of domestic animals, the llamas must be kept principally for the fleece, just as sheep are in some places, where mutton is not esteemed for food, since they would not be valuable as beasts of burden, except in very mountainous districts; and their flesh, although eaten in South America, is not esteemed by such of our countrymen as have tasted it. Lieut. Phelps says of it: 'I have tried the flesh, and, although not partial to it, could live upon it if hard pushed.' That observation was made of the animal in its wild state, called in Chili, Guanaco, but generally believed to be the original of the Peruvian llama, alpaca, or vicuna. The latter name is sometimes applied as the generic term of the race, and the other two names those of varieties differing no more than varieties of sheep, the Alpacas being considerably smaller than the others, and more woolly; some of the variety called llamas are as tall as good sized yearling bovines.

In a wild state, the guanacos inhabit the chain of mountains from Terra del Fuego to the Cordilleras in Peru, choosing their pasture ground just below the snow line. In a domestic state the llamas are used as beasts of burden, in mountain or plain, in cold and heat, often traversing snowy altitudes and tropical valleys upon the same journey. They are much used in the Andes, by the miners, to bring down ore and carry up supplies; travelling twelve or fifteen miles a day, with loads of 100 lbs. each, and living upon the coarsest and most scanty supply of herbage, and, like the camel, enduring days of toil without water. It is possible the llama, as well as the camel, may yet be used to advantage by travellers upon our great American deserts. It is stated in the history of Peru, that troops of llamas, a thousand in number, used to be common, all bearing their loads, and travelling under the guidance of a few men, over regions where no other beast could obtain a footing.

"One of the herd offered for sale on Saturday, was exhibited loaded with packs, as he would be upon a journey. All of them, even a lamb of a few months old, are broken to the halter, and are very docile and tractable. Their countenances exhibit marked expressions of intelligence—the eyes are very bright and sight keen. The colour is generally that of brown or black sheep—some of them pretty nearly jet black. Some of the males are grey, or nearly white, with white faces. The shape of the head, face, ears, and neck is like that of a native sheep, except the neck is more elongated. The cloven hoofs are larger, and the legs longer than the tallest sheep, and the bodies though longer, do not appear much larger than some of the tall varieties of sheep. The anatomy is curious in this, that the thigh seems to proceed from the hip-joint, but with little connection with the body.

"The fleece is from four to six inches long, fine and soft within, with coarse hairs thinly scattered through it, and projecting beyond the mass. It very much resembles the fleece of a black sheep. We should judge the average weight of fleeces might be about ten pounds—the bellies being generally bare—and the value is greater than that of wool. The excellence and durability of alpaca cloth are generally known."

The report went on to say that the sale was not successful, the biddings never reaching the reserved price of 100 dollars each animal.

He (Mr. Gee) subsequently bought the animals, and took them by the steamer, *City of New York*, to Glasgow, where they arrived shortly before the agricultural show in that city. He exhibited them at a charge of sixpence each, and had as many as 1,200 visitors. He afterwards brought them to Birmingham, at the time of the Queen's State visit, and although there were about a million of people congregated on that occasion, not more than twenty-eight persons honored his alpacas with a visit. The people seemed to have no idea what the animals were. Then he brought them to London, in the hot weather of June. He sold three to Mr. Pattison, two to Miss Coutts, and ten to Mr. G. A. Lloyd, the latter at £25 each, which paid him very well, but they afterwards sold for £60 each. Mr. E. Wilson, the well-known editor of the Melbourne *Argus*, took the remainder of the flock for £700, which was at the rate of £23 per head. During the time the flock was in his possession, they grazed at Acton, about five miles from London, and they got quite fat. They started for Australia, and like most other emigrants, they did even better there. With regard to the acclimatisation of these animals, it could only be proved by time. In the case of the Angora goat, which was introduced into Australia some years ago, a very beautiful fleece was produced when they got the real thing itself, but even amongst the goats brought from Turkey there was a white streak which spoiled its appearance, and the only doubt was that the skins would run "kempy," and that in the course of a few generations the fleece would become as coarse as the hair of the common goat. He did not, however, think that would be the case with the alpaca, although the fleece had not the oiliness of the Merino sheep, and from the nature of the food in Australia, he believed the wool would be improved in texture. With regard to the suitability of the climate there, and the supply of proper food for these animals, there could be no doubt. They might walk for hundreds of miles up to their knees in grass, which was excellent food, and the animals would eat it when it was in a dry state. At the time he bought his flock in New York, he was wholly ignorant of what was the proper food for them; but during the passage he fed them with Indian maize, which they readily ate, as also hay, and they were in better condition after thirteen days sea passage than when they were put on board. There was one point which he felt some delicacy in touching upon. He was glad to hear from Mr. Macarthur that Mr. Ledger was likely to reap some reward for his exertions; but with regard to himself, he was somewhat in the position of the Irish ostler, who, when a gentleman was about to drive off without having handed him the customary gratuity, called after him, "Please yer honor, if my master asks me what you gave the ostler, what am I to tell him?" and if any one asked him what he had got, his reply would be—nothing.

Mr. DAVIS would be sorry to divert the discussion from the congratulatory tone which had characterised it, for it appeared they were a "Mutual Congratulation Society" that evening. One thing they might certainly congratulate themselves upon—that was upon the perseverance and energy which had pre-eminently distinguished our countryman, Mr. Ledger, in this matter. All who had taken any interest in affairs in Australia and South America, must be aware that the difficulties, dangers, and privations which Mr. Ledger had gone through were such as few men would have the courage to undergo, even with the prospect of success in view. Yet whilst he tendered to Mr. Ledger the full measure of gratitude for what he had done, he could not look upon this matter from the *couleur de rose* point of view in which some regarded it. If they looked at the locality in which these animals originated—they lived in extremely mountainous regions, almost within the reach of perpetual snow. The vicuna lived in the highest regions of all. In Australia they heard of snowy ranges, but those were mountains which were covered with snow only during certain portions of the year. Snow did not exist there over any great extent of country for any great length of time, and in such localities as it most existed the country was rather barren and rocky. It was true these animals could live upon hard fare, but they could not live upon mere rocks. The Snowy Mountains of Australia were, to a great extent, of a rocky nature, and the quantity of grass was small. They had been told of the country explored by Mr. Stuart, but it should at the same time have been stated that it was a country of intense heat, and these were not animals to stand great heat, but were rather adapted for cold climates. No doubt they would live up to their bellies in snow, but he for one did not look forward with the hope that some did to the propagation of this animal in Australia. He would rather point to this subject as affording to enterprising young men in this country an opening to make their fortunes, by going to Peru and Bolivia, and promoting the growth of wool there from the animal in its native climate. It might be said that South America was a badly governed country, but there were Englishmen making money there very fast. He quite approved of the course which had now been taken by the Government of Australia. They were now disposed to do what they should have done in the first instance. They might now say to Mr. Ledger

"By your energy and perseverance you have proved yourself a benefactor to your country; here is the amount you have spent; take back your animals; increase and prosper. If you succeed, this country must benefit thereby, but in any case it is no use for the Government to turn sheep farmers." As to introducing these animals into England, he thought it would be unwise. At present they had animals which produced both wool and mutton, and to bring into a country like this animals which produced fleeces only, would never answer. He was not afraid of this being attempted, for the very best reason with Englishmen—that it would not pay. They wanted an animal which gave them both clothes and food, and this they had in the sheep, from the Scotch sheep which dwelt in the snows of the north, to the South Down, which flourished in the mildest climate and on the dryest lands. He could not sit down without expressing his deep obligation to Mr. Charles Ledger for what he had done, and all that he had contemplated doing for Australia. He regarded him as a great benefactor to that country if he did not succeed, but if he succeeded he would be an enormous benefactor both to Australia and his fellow-countrymen at home.

Dr. CRISP believed that there was an error as to the date at which these animals were first introduced into Europe. He understood it to be stated in the paper that it was not until the year 1815 that the llama was first introduced, but he believed it would be found that specimens of these animals were brought to the Jardin des Plantes, in Paris, as early as 1808. He begged to add his meed of praise to Mr. Ledger for the exertions he had made in this matter, and to the gentleman who had so ably brought this interesting subject before them.

The CHAIRMAN, in closing the discussion, thought the meeting would agree with him that, if there was one paper more than another which was particularly within the province of the society, it was the one they had just heard. The subject of how to improve the cultivation, and increase the production of so important a staple as wool, was one well worthy of their most attentive consideration. There was nothing in the paper of a speculative character. They had a collection of facts and details upon an important subject which might be turned to good account, if not at the present moment, certainly at some future time. It was their duty, as a society, to give every encouragement to gentlemen who came before them with such valuable information as Mr. Ledger had collected—not in the ordinary sense of the term, but received directly from his brother, who had devoted so many years of his life to the promotion of this object.

His friend on the left (Mr. Davis) had characterised the meeting as one of mutual congratulation. He (the Chairman) thought it was justly so. At the same time, his friend had advanced opinions somewhat at variance with those expressed in the paper. That was one true object of discussion. They did not ask gentlemen to approve of everything that was said. They always invited discussion upon what was advanced in the paper, and the greater the variety of opinion that was elicited the more valuable the meetings of the society became. They had heard a very remarkable fact that evening, which showed how important a part the commerce of this country played in the industry of the world. Englishmen, in the first instance, had been the means of introducing the animal spoken of from South America into Australia, and then they afforded the best market to which the commodity produced could be sent. That was a noteworthy instance of the influence of British commerce all over the world. He was sure they would all feel indebted to the gentleman who had brought such a collection of facts before them in so clear a manner as had been done that evening, and that they would cordially concur in a vote of thanks to Mr. Ledger for his paper.

The vote of thanks having been passed,

Mr. LEDGER, in reply on the discussion, said he had studiously avoided introducing into this subject matters of a purely personal or pecuniary character; he had not spoken of the dangers and difficulties his brother had undergone more than was essential to the proper elucidation of the subject; but as remuneration had been alluded to, he fully concurred in the hope that his brother would ultimately reap a large reward for the accomplishment of his enterprise. This depended entirely, however, upon the success of the animals. He had heard (not from his brother), that at Copiapo £42,000 had been offered for the flock for the French Government, and declined. He thought the New South Wales Government had acted with forethought and generosity in giving his brother £15,000, every farthing of which, however, had been remitted to South America, to liquidate liabilities incurred in the prosecution of the enterprise, leaving a balance still due of £1,080, and he was happy to say his brother had concluded an arrangement with the Colonial Government, which was waiting the sanction of the Legislature, to pay him cash £2,000, and £3,000 in annual payments of £500 for six years, to provide him pasture grounds free of rent or taxes for 12 years; at the expiration of which the animals were to be his property on payment of the principal of £20,000 without interest. The produce of the wool in the twelve years was to be his, but he was not to be permitted to sell any of the animals without the consent of the Colonial Government.

The paper was illustrated by a collection of alpaca skins, contributed by Mr. Murietta; vicuna skins, by Mr. Skinner Row; specimens of alpaca wool, by Mr. Gee; of llama wool, by Miss Burdett Coutts; and pieces of alpaca fabric by Messrs. Edwards.

5 JA 70

www.ingramcontent.com/pod-product-compliance
Lightning Source LLC
Chambersburg PA
CBHW081123240526
45470CB00019B/2927